Breathe Easy by the Fire:

A Guide to Safe Wood Stove Use for Respiratory Health

Combining Traditional Heating with

Modern Wellness Practices for a Healthier Home Environment

Alessio rocchl

Table of Contents

Introduction: The Warmth of Wellness

This introduction highlights the connection between traditional wood stove heating and modern respiratory health concerns. It sets the stage for understanding how these two seemingly unrelated topics intersect and why this guide is essential for creating a healthy home environment.

Chapter 1: The Resurgence of Wood Stoves

Explores the growing popularity of wood stoves as a sustainable heating solution. This chapter discusses their eco-friendly benefits, cost savings, and charm, while also addressing common misconceptions about wood stove safety and pollution.

Chapter 2: The Science of Clean Air

Introduces the basics of respiratory health, including how air quality impacts breathing and overall wellness. This chapter explains why clean indoor air is vital and how wood stoves, when used properly, can coexist with good air quality.

Chapter 3: Choosing the Right Wood Stove

A buyer's guide to selecting the best wood stove for your home. This chapter covers factors such as size, efficiency, emissions ratings, and design considerations, ensuring your choice aligns with your heating needs and health priorities.

Chapter 4: Proper Installation for Safety and Efficiency

Details the steps for safely installing a wood stove, including proper venting and positioning. It emphasizes how a well-installed stove reduces emissions and improves air circulation, creating a healthier indoor environment.

Chapter 5: Fueling the Fire: Choosing the Right Wood

Discusses the types of wood that burn cleanly and efficiently. Readers will learn how to identify, season, and store firewood to minimize indoor air pollution and maximize heat output.

Chapter 6: Maintaining Your Wood Stove for Longevity and Health

Covers routine maintenance practices, such as chimney cleaning and ash disposal, to ensure safe operation and reduced emissions. This chapter highlights how proper upkeep directly supports respiratory health.

Chapter 7: Understanding Indoor Air Pollution

Explains how wood smoke can affect indoor air quality and health if not properly managed. This chapter introduces practical solutions, including air filtration systems and ventilation strategies, to mitigate risks.

Chapter 8: The Role of Portable Nebulizers in Respiratory Health

Provides an overview of portable nebulizers, their functions, and how they can offer relief for those with respiratory conditions. This chapter explains why having one on hand is a smart complement to using a wood stove.

Chapter 9: Creating a Healthy Home Environment

Offers tips on designing a home that promotes both warmth and wellness. This chapter integrates the use of wood stoves, air purifiers, humidifiers, and other tools to create a balanced and health-conscious living space.

Chapter 10: Emergency Preparedness with Wood Stoves

Discusses the role of wood stoves during power outages and other emergencies. This chapter focuses on maintaining safety while relying on wood stoves for heating during challenging times, with a particular emphasis on health considerations.

Chapter 11: Stories from Warm, Healthy Homes

Includes real-life anecdotes and testimonials from families who have successfully combined wood stove heating with respiratory health strategies. These stories inspire and validate the practical advice offered in the book.

Chapter 12: Frequently Asked Questions

Addresses common concerns and misconceptions about wood stoves and respiratory health. This chapter provides concise, actionable answers to reader questions for quick reference.

Conclusion: Warmth, Health, and Harmony

Summarizes the key takeaways from the book, reinforcing the idea that warmth and respiratory health can coexist with proper knowledge and practices. Encourages readers to take the first steps toward a healthier, cozier home.

Introduction: The Warmth of Wellness

There's something undeniably primal about the crackling of a fire. For thousands of years, humans have gathered around flames, not just for warmth but for connection, comfort, and survival. A wood stove, often the heart of a home, serves as a bridge between past and present—a reminder of simpler times when life revolved around the hearth. But in modern living, where convenience often trumps tradition, the wood stove stands as both a nostalgic relic and a practical tool. Yet, there's a question many people don't think to ask when they light a fire: *How does this choice affect the air I breathe and the health of my family?*

In an era when health and wellness dominate public discourse, the intersection between our choices for sustainable living and personal well-being takes center stage. This book explores how you can embrace the warmth and charm of a wood stove while safeguarding your health, particularly your respiratory system, from the risks associated with indoor air pollution. It's not about sacrificing comfort for safety or vice versa—it's about balance.

A Return to Simplicity

Modern society is witnessing a renewed interest in practices that bring us closer to nature. From the rise of urban gardening to the popularity of homesteading, people are searching for ways to feel connected to their roots. The wood stove embodies this desire perfectly, offering a sustainable, off-grid heating solution that's both cost-effective and aesthetically appealing. For those seeking to reduce their carbon footprint, the wood stove represents a tangible step toward greener living. Yet, this romanticized image of a wood stove—its flames dancing in the hearth, its warmth spreading through a room—often leaves out a critical aspect: the potential impact on the air quality inside your home.

While wood stoves have a lower carbon footprint compared to fossil fuel-based heating systems, they are not without challenges. The smoke they emit, even in small amounts, can introduce fine particulate matter (PM2.5) into your living space. These microscopic particles can infiltrate your respiratory system, triggering health issues if not managed properly. The key is not to avoid the use of a wood stove entirely, but to understand how to use it wisely and safely.

The Science of Breathing: Why Indoor Air Matters

Breathing is so automatic that most of us rarely stop to consider the quality of the air filling our lungs. Yet, according to the Environmental Protection Agency (EPA), the air inside our homes can be up to five times more polluted than outdoor air. This statistic is alarming, especially when you consider how much time we spend indoors—whether we're working from home, relaxing, or sleeping.

For those with pre-existing conditions like asthma or chronic obstructive pulmonary disease (COPD), poor indoor air quality can exacerbate symptoms. Even healthy individuals can experience short-term effects such as irritation of the eyes, nose, and throat, or long-term impacts on lung function and cardiovascular health. When you add a wood stove into the equation, the stakes become even higher. The good news? With the right strategies, you can enjoy the benefits of a wood stove without compromising your health.

The Emotional Comfort of a Fire

There's a reason why the concept of "hygge" (a Danish word that loosely translates to a sense of coziness and well-being) often includes imagery of a fire. The flicker of flames, the gentle crackle of wood, and the enveloping warmth evoke feelings of safety and relaxation. Studies have shown that the mere presence of a fire can reduce stress and lower blood pressure, reinforcing its role as a natural source of comfort.

This emotional connection to fire is deeply ingrained in our psychology. For millennia, fire has been central to human survival. It provided light in the dark, warmth in the cold, and a means to cook food. In a modern context, a wood stove brings these primal benefits into the home, fostering a sense of grounding in an otherwise chaotic world.

However, there's a flip side to this comfort. For those who rely heavily on wood stoves, especially in off-grid or rural settings, the improper use of a stove can lead to air quality issues that negate the health benefits of the soothing fire. The challenge, then, is to find a way to retain the emotional and physical comfort of a wood stove while addressing its potential risks.

Bridging Tradition and Technology

One of the most exciting developments in the wood stove industry is the integration of modern technology. Advanced stove designs now include catalytic and non-catalytic systems that drastically reduce emissions. These innovations allow you to enjoy the benefits of a wood stove without the drawbacks of older models, which often lacked the efficiency and safety features of their modern counterparts.

But technology isn't limited to the stoves themselves. Homeowners are increasingly turning to indoor air quality monitors to keep tabs on particulate levels and air purifiers to remove airborne pollutants. Portable nebulizers, often associated with respiratory therapy, have also found a place in homes with wood stoves, offering a convenient way to manage any respiratory discomfort caused by occasional exposure to smoke.

The integration of these tools is not just a matter of convenience—it's a way to make wood stove heating a viable option for families who prioritize both sustainability and health. The message is clear: you don't have to choose between tradition and progress; you can embrace both.

Sustainability Meets Wellness

At its core, the use of a wood stove is a sustainable choice. Wood is a renewable resource, and when burned efficiently, it releases the same amount of carbon dioxide as it would if it were left to decompose naturally. This makes it an appealing option for those looking to reduce their reliance on fossil fuels.

However, the concept of sustainability must extend beyond environmental concerns to include personal wellness. True sustainability means creating a home environment that supports your health while minimizing your impact on the planet. This holistic approach requires an understanding of how to balance the practicalities of wood stove use with strategies to maintain clean air and optimal respiratory health.

The Hidden Costs of Neglecting Air Quality

For many, the decision to use a wood stove is driven by financial considerations. Heating a home with wood is often more affordable than relying on electricity or gas, especially in areas where wood is readily available. However, the hidden costs of neglecting air quality can far outweigh these savings.

Medical expenses, lost productivity due to illness, and even the emotional toll of managing health issues related to poor air quality can add up over time. By taking proactive steps to ensure the safe use of a wood stove, you're not only protecting your health but also preserving the financial benefits of this heating method.

A New Perspective on Home Heating

The modern home is a complex ecosystem, where every choice—from the materials used in construction to the appliances selected—affects the overall quality of life for its inhabitants. The wood stove, once a symbol of simplicity, now represents a nuanced decision that requires careful consideration.

This book aims to empower you with the knowledge and tools

needed to make informed decisions about wood stove use. From selecting the right model to understanding how it interacts with your home's air quality, this guide will walk you through every step of the process. Along the way, you'll discover how to combine the best of traditional practices with modern innovations to create a home that's both warm and healthy.

A Journey Worth Taking

As you turn the pages of this book, you'll embark on a journey that goes beyond the practicalities of heating your home. You'll explore the science of air quality, the art of creating a cozy living space, and the joy of finding harmony between tradition and wellness. Whether you're a seasoned wood stove user or someone considering one for the first time, this guide will equip you with the insights and inspiration needed to embrace the warmth of wellness in your home.

Chapter 1: The Resurgence of Wood Stoves

In an era dominated by sleek, high-tech appliances and modern heating systems, the wood stove stands out as a paradoxical champion of simplicity and tradition. It's a symbol of both nostalgia and sustainability, an artifact of the past that has found new relevance in today's eco-conscious world. But why, in a time of unprecedented technological advancement, are so many individuals turning back to the warmth of wood stoves? To understand this resurgence, we must explore the intertwining threads of environmental awareness, cultural shifts, and practical necessity.

A Historical Perspective: From Necessity to Nostalgia

The story of the wood stove begins centuries ago when it was not a choice but a necessity. In pre-industrial societies, wood stoves provided the most reliable means of heating homes and cooking food. Early designs were rudimentary, often nothing more than iron boxes with basic venting systems. Over time, innovations in design and materials transformed these crude stoves into more efficient and aesthetically pleasing fixtures of domestic life.

By the mid-20th century, however, wood stoves began to fade from prominence. The rise of centralized heating systems powered by oil, gas, and electricity offered convenience and consistency that wood stoves could not match. Urbanization also played a role, as smaller living spaces and stricter building codes made wood stove installation less practical in city apartments.

Despite this decline, wood stoves never disappeared entirely. For rural and off-grid communities, they remained indispensable. For others, they became cherished relics—symbols of a simpler, more self-reliant way of life. As modern challenges such as climate change and economic instability emerged, this nostalgia began to fuel a new appreciation for wood stoves.

Environmental Awareness: A Key Driver

One of the most significant factors driving the resurgence of wood stoves is the growing awareness of environmental issues. Unlike fossil fuels, wood is a renewable resource. When sourced sustainably, it offers a heating option that can reduce reliance on non-renewable energy sources. Moreover, the carbon released during the burning of wood is part of a natural cycle; it's the same amount of carbon that would be released if the wood were left to decompose naturally.

This concept, known as carbon neutrality, appeals to eco-conscious consumers. However, it's essential to note that not all wood stove use is created equal. Modern, EPA-certified stoves are designed to burn wood more efficiently, producing less smoke and fewer particulates compared to older models. This efficiency not only reduces emissions but also makes better use of the wood, providing more heat per log.

Economic Factors: Affordable and Reliable Heating

For many, the decision to use a wood stove is driven by economics. Heating costs can represent a significant portion of household expenses, particularly in colder climates. Wood, especially when harvested locally, is often more affordable than oil, gas, or electricity. Additionally, wood stoves provide a level of energy independence that appeals to those wary of fluctuating utility prices or potential power outages.

This economic advantage becomes even more pronounced during times of crisis. Whether it's a winter storm that knocks out electricity or a geopolitical event that disrupts fuel supplies, a wood stove offers a reliable source of heat. This resilience is particularly valued in rural areas where access to alternative heating options may be limited.

The Aesthetic and Emotional Appeal

Beyond practicality, wood stoves possess an undeniable aesthetic and emotional appeal. There's something deeply satisfying about the sight and sound of a crackling fire. It creates an atmosphere of warmth and coziness that modern heating systems simply cannot replicate. In Scandinavian countries, this concept is often referred to as "hygge"—a sense of comfort and well-being associated with simple, pleasurable experiences.

Wood stoves also serve as focal points in home design. Manufacturers have recognized this, offering a range of styles from rustic to contemporary. Whether it's a cast-iron stove with intricate detailing or a sleek, minimalist model, the wood stove has become as much a statement piece as a functional appliance.

Technological Advancements: Making Wood Stoves Smarter and Safer

The wood stove of today is a far cry from its historical predecessors. Technological advancements have addressed many of the concerns that once plagued wood stove use, such as inefficiency and pollution. Modern stoves feature innovations like secondary combustion systems, which burn off excess gases to increase efficiency and reduce emissions. Catalytic converters, similar to those used in cars, further minimize pollutants.

Smart technology has also made its way into the world of wood stoves. Some models now include digital controls that allow users to monitor and adjust temperatures with precision. Others are equipped with sensors that provide real-time data on particulate levels, helping users maintain a safe and clean-burning fire. These advancements make wood stoves more accessible to a broader audience, including those who may have previously been deterred by their perceived complexity.

Cultural Shifts: A Desire for Self-Sufficiency

The resurgence of wood stoves also reflects broader cultural trends toward self-sufficiency and sustainability. Movements like homesteading and off-grid living have gained popularity, driven by a desire to reclaim control over essential aspects of life such as food, water, and energy. For these communities, the wood stove is more than just a heating device; it's a symbol of independence.

Even among those who don't fully embrace off-grid living, the wood stove represents a step toward greater self-reliance. Learning to chop, stack, and season wood is a tangible way to engage with the natural world and take responsibility for one's own energy needs. This hands-on approach resonates in an age where many aspects of daily life feel increasingly automated and disconnected from nature.

Regulatory Changes and Incentives

Government policies have also played a role in the wood stove's comeback. In many regions, older, inefficient stoves are being phased out in favor of newer, cleaner models. Programs offering tax credits or rebates for the purchase of EPA-certified stoves have made it more financially feasible for households to upgrade.

These regulations not only reduce the environmental impact of wood stove use but also contribute to public health by minimizing air pollution. For consumers, the availability of financial incentives adds another layer of appeal to an already cost-effective heating option.

Challenges and Considerations

While the resurgence of wood stoves is largely positive, it's not without challenges. Improper use or installation can negate many of the benefits, leading to issues such as inefficient burning, increased emissions, or even safety hazards. Education is critical to ensuring that users understand how to operate and maintain their stoves effectively.

There's also the issue of wood sourcing. For wood stoves to be a truly sustainable option, the wood must be harvested responsibly. Overharvesting can lead to deforestation and other environmental problems, undermining the ecological benefits of wood stove use. Encouraging the use of sustainably sourced or locally harvested wood is an essential part of promoting the responsible use of this technology.

The Role of Community and Tradition

Finally, it's worth noting the communal aspect of wood stove culture. In many rural areas, the process of chopping, stacking, and sharing wood is a tradition passed down through generations. It's a communal activity that fosters connection and a sense of shared responsibility. For urban dwellers, joining a wood co-op or sourcing wood from local suppliers can create similar opportunities for community engagement.

As wood stoves continue to gain popularity, they are not just heating homes; they are rekindling connections—to the past, to the environment, and to each other. This chapter has outlined the various factors contributing to their resurgence, setting the stage for deeper exploration in the chapters to come.

Chapter 2: The Science of Clean Air

The air we breathe is as vital to life as water or food, yet it often goes unnoticed until it's compromised. Our respiratory systems are designed to extract oxygen from the atmosphere, filtering out impurities to keep our bodies functioning optimally. However, the quality of the air around us can vary significantly, influencing everything from our short-term well-being to long-term health outcomes. This chapter dives deep into the science of clean air, exploring its components, how it interacts with our respiratory systems, and why maintaining clean air in your home—especially if you use a wood stove—is so crucial.

The Composition of Air

At its most basic level, air is a mixture of gases. The Earth's atmosphere is composed of approximately:

- 78% nitrogen (N_2)
- 21% oxygen (O_2)
- 0.93% argon (Ar)
- 0.04% carbon dioxide (CO_2)
- Trace amounts of other gases, including neon, helium, methane, and ozone

While these primary components are largely stable, the air we breathe also contains variable components such as water vapor, particulate matter, and volatile organic compounds (VOCs). These additional elements, especially in higher concentrations, can impact air quality and pose health risks.

Indoor vs. Outdoor Air Quality

When most people think of air pollution, they envision smog-filled cityscapes or industrial smoke stacks. However, studies by the Environmental Protection Agency (EPA) reveal that indoor air can be two to five times more polluted than outdoor air. In some cases, particularly in homes with poor ventilation, it can be even worse.

Indoor air pollutants fall into two categories:

1. **Particulate Matter (PM)**: Tiny particles suspended in the air that can include dust, soot, and smoke. PM2.5— particles with a diameter of 2.5 micrometers or smaller— is especially concerning because it can penetrate deep into the lungs and even enter the bloodstream.
2. **Gaseous Pollutants**: These include carbon monoxide (CO), nitrogen dioxide (NO_2), and VOCs, which are emitted by household products, furniture, and certain heating systems.

Wood stoves, while efficient and sustainable in many ways, can contribute to both types of indoor air pollution if not managed properly. Understanding the science of these pollutants is the first step toward mitigating their effects.

The Respiratory System and Air Quality

To appreciate the importance of clean air, it's essential to understand how the respiratory system functions. When we inhale, air travels through the nose or mouth, down the trachea, and into the lungs via the bronchi. Within the lungs, it reaches tiny air sacs called alveoli, where oxygen is exchanged for carbon dioxide in the blood.

This process is incredibly efficient, but it's also delicate. The respiratory system relies on clean air to function optimally. When pollutants like PM2.5 or VOCs enter the lungs, they can irritate the respiratory tract, trigger inflammation, and even damage lung tissue. Over time, exposure to poor air quality can contribute to chronic respiratory conditions such as asthma, bronchitis, and emphysema.

Common Indoor Air Pollutants and Their Effects

Several pollutants can compromise indoor air quality, each with its own sources and health implications:

1. **Particulate Matter (PM2.5 and PM10):**
    - **Sources**: Wood stoves, tobacco smoke, cooking, and outdoor air infiltration.
    - **Health Effects**: Can cause respiratory irritation, reduced lung function, and cardiovascular issues.
2. **Carbon Monoxide (CO):**
    - **Sources**: Incomplete combustion from wood stoves, gas appliances, and vehicles in attached garages.

- **Health Effects**: At high levels, it's a deadly gas that interferes with oxygen delivery in the body, leading to headaches, dizziness, and, in severe cases, death.

3. **Volatile Organic Compounds (VOCs)**:
   - **Sources**: Cleaning products, paints, varnishes, and emissions from burning wood.
   - **Health Effects**: Can cause short-term symptoms like headaches and eye irritation, with long-term exposure linked to liver and kidney damage.

4. **Nitrogen Dioxide (NO_2)**:
   - **Sources**: Combustion processes, including those in wood stoves and gas appliances.
   - **Health Effects**: Can irritate airways and exacerbate respiratory conditions like asthma.

5. **Mold Spores**:
   - **Sources**: Damp areas in the home, particularly bathrooms and basements.
   - **Health Effects**: Can cause allergic reactions, respiratory issues, and infections in immunocompromised individuals.

Wood Stoves and Air Quality

Wood stoves are a double-edged sword when it comes to air quality. On one hand, they're an efficient, renewable source of heat. On the other, they can release pollutants into the air if not properly maintained or operated. Key factors that influence the impact of a wood stove on indoor air quality include:

1. **Stove Design**: Modern, EPA-certified wood stoves are designed to burn wood more completely, reducing emissions.
2. **Wood Quality**: Using seasoned, dry wood minimizes smoke and particulates.
3. **Ventilation**: Properly installed chimneys and flues ensure that smoke is vented outside rather than accumulating indoors.
4. **Maintenance**: Regular cleaning of the stove and chimney prevents buildup that can lead to inefficient burning and increased emissions.

The Role of Ventilation

Ventilation is one of the most critical elements in maintaining clean indoor air. In homes with wood stoves, adequate ventilation ensures that pollutants are effectively removed and replaced with fresh outdoor air. This can be achieved through:

1. **Mechanical Ventilation**: Systems like HRVs (Heat Recovery Ventilators) and ERVs (Energy Recovery Ventilators) exchange indoor air with outdoor air while maintaining energy efficiency.
2. **Natural Ventilation**: Opening windows and doors to allow cross-ventilation.
3. **Exhaust Fans**: Installing fans in kitchens and bathrooms to remove localized pollutants.

Air Purification and Filtration

In addition to ventilation, air purification and filtration systems can play a vital role in maintaining clean air. These systems work by removing particulates and pollutants from the air, making them particularly useful in homes with wood stoves. Options include:

1. **HEPA Filters**: High-Efficiency Particulate Air (HEPA) filters capture 99.97% of particles as small as 0.3 microns, making them highly effective against PM2.5.
2. **Activated Carbon Filters**: These filters absorb VOCs and odors, complementing HEPA filtration.
3. **UV-C Air Purifiers**: Use ultraviolet light to kill bacteria, viruses, and mold spores.

Behavioral Strategies for Cleaner Air

While technology plays a crucial role, individual behavior can also significantly impact indoor air quality. Strategies include:

1. **Burning the Right Wood**: Only use seasoned hardwoods to minimize smoke and emissions.
2. **Avoiding Overloading**: Overfilling the stove can lead to incomplete combustion.
3. **Regular Maintenance**: Clean the stove, chimney, and filters regularly to ensure efficient operation.
4. **Reducing Additional Pollutants**: Limit the use of candles, incense, and chemical cleaning products.

Health Benefits of Clean Air

The benefits of maintaining clean air extend beyond avoiding health issues. Clean air can improve sleep quality, boost cognitive function, and enhance overall well-being. For families with children, it's especially important, as young lungs are more vulnerable to pollutants.

Research has also shown that improving air quality can reduce the risk of developing chronic conditions such as heart disease and diabetes. In essence, investing in clean air is an investment in long-term health.

Emerging Research and Technologies

The field of air quality research is continually evolving, with new technologies and insights emerging regularly. Some promising developments include:

1. **Low-Emission Stoves**: Advances in wood stove design are making them cleaner and more efficient than ever.
2. **Real-Time Air Quality Monitoring**: Devices that provide immediate feedback on indoor air quality, empowering homeowners to take action.
3. **Smart Home Integration**: Systems that connect air purifiers, ventilation, and heating systems for seamless management.

These innovations are making it easier than ever to maintain clean air, even in homes with wood stoves.

Chapter 3: Choosing the Right Wood Stove

Selecting the right wood stove for your home is not just about picking a heating appliance—it's about finding a companion that complements your lifestyle, aligns with your values, and enhances your living environment. A wood stove is an investment, not only in warmth but also in sustainability, design, and safety. This chapter will guide you through the factors to consider when choosing a wood stove, ensuring that your choice aligns with your heating needs, aesthetic preferences, and environmental goals.

Understanding Your Heating Needs

The first step in selecting the right wood stove is understanding your specific heating requirements. A wood stove that's too small won't adequately heat your space, while one that's too large can waste fuel and create an uncomfortably hot environment. Here are the key considerations:

1. **Room Size**: Measure the square footage of the area you want to heat. As a general rule, a stove with a heating capacity of 25 to 30 BTUs per square foot is sufficient for most homes. For example, a 1,000-square-foot space would require a stove rated at 25,000 to 30,000 BTUs.
2. **Climate**: Consider the climate in your region. Colder climates may require a more robust heating system, while milder areas might be well-served by a smaller stove.
3. **Home Insulation**: Evaluate the insulation in your home. Poorly insulated homes lose heat quickly, necessitating a more powerful stove. Modern, well-insulated homes can retain heat more effectively, allowing for smaller, more efficient stoves.
4. **Primary vs. Supplemental Heating**: Determine whether the stove will serve as your primary heat source or supplement an existing system. A supplemental stove may not need as much capacity as a primary heat source.

Types of Wood Stoves

Wood stoves come in a variety of styles and configurations, each with its own advantages and considerations. Understanding the different types will help you narrow down your options:

1. **Catalytic Wood Stoves**:
   - **How They Work**: Catalytic stoves use a catalytic combustor to re-burn gases and particles, achieving a cleaner and more efficient burn.
   - **Advantages**: High efficiency, reduced emissions, and longer burn times.
   - **Considerations**: Higher upfront cost and the need to replace the catalytic combustor every 5 to 10 years.
2. **Non-Catalytic Wood Stoves**:
   - **How They Work**: These stoves rely on a series of baffles to direct air flow and promote secondary combustion.
   - **Advantages**: Lower initial cost, simpler operation, and less maintenance.
   - **Considerations**: Slightly less efficient and shorter burn times compared to catalytic models.
3. **Hybrid Wood Stoves**:
   - **How They Work**: Combine catalytic and non-catalytic technologies for maximum efficiency and flexibility.
   - **Advantages**: Best of both worlds with high efficiency and lower emissions.

- **Considerations**: Higher cost and slightly more complex maintenance.

4. **Pellet Stoves**:
    - **How They Work**: Burn compressed wood pellets rather than traditional logs, offering automated feeding systems for convenience.
    - **Advantages**: High efficiency, minimal emissions, and easy operation.
    - **Considerations**: Requires electricity to operate and access to pellet supplies.

Materials and Construction

The material of your wood stove plays a significant role in its performance and aesthetics. Common materials include:

1. **Cast Iron**:
    - **Advantages**: Retains heat for a long time after the fire has burned out, offering consistent warmth.
    - **Design**: Often features intricate designs and a traditional appearance.
    - **Considerations**: Requires periodic maintenance to prevent cracking.
2. **Steel**:
    - **Advantages**: Heats up quickly and is generally

more affordable than cast iron.
    - **Design**: Sleek, modern appearance with fewer design flourishes.
    - **Considerations**: Cools down faster than cast iron.
3. **Soapstone**:
    - **Advantages**: Exceptional heat retention and a unique, natural aesthetic.
    - **Design**: Distinctive look that complements rustic and contemporary styles.
    - **Considerations**: Higher cost and heavier weight, which may complicate installation.
4. **Ceramic**:
    - **Advantages**: Offers excellent heat retention and aesthetic versatility.
    - **Design**: Available in a variety of colors and patterns.
    - **Considerations**: Fragility compared to other materials.

EPA Certification and Efficiency

Environmental considerations are increasingly important for modern homeowners. An EPA-certified wood stove ensures compliance with strict emission standards, making it a cleaner choice for both your home and the planet. Here's what to look for:

1. **Emission Limits**: EPA-certified stoves emit less than 2.5 grams of particulate matter per hour, compared to older models that can release over 30 grams per hour.
2. **Efficiency Ratings**: Look for stoves with efficiency ratings of 70% or higher. A more efficient stove uses less wood and produces less waste.
3. **Tax Incentives**: Many EPA-certified stoves qualify for government rebates or tax credits, offsetting the initial cost.

Design and Aesthetics

Your wood stove isn't just a heating appliance—it's also a central element of your home's design. Whether you prefer a rustic, traditional look or a sleek, modern aesthetic, there's a wood stove to suit your style. Consider the following:

1. **Size and Scale**: Choose a stove that fits proportionally within your space. A large stove in a small room can feel overwhelming, while a tiny stove in a large space might look out of place.
2. **Finish**: Stoves are available in a variety of finishes, from classic black cast iron to enameled surfaces in vibrant

colors.

3. **Viewing Window**: Many modern stoves feature large, heat-resistant glass panels that showcase the fire, adding ambiance to your space.

Installation Considerations

Proper installation is critical for both safety and performance. Before purchasing a stove, evaluate the following:

1. **Chimney and Venting**: Ensure you have an appropriate chimney or flue system. Existing chimneys may need to be lined or upgraded to meet modern safety standards.
2. **Clearances**: Wood stoves require specific distances from walls and combustible materials. Check the manufacturer's guidelines to ensure compliance.
3. **Floor Protection**: Stoves must be placed on a fireproof surface, such as a hearth pad, to protect your flooring from heat and sparks.
4. **Professional Installation**: While some homeowners opt for DIY installation, hiring a certified professional ensures that your stove is installed safely and meets local building codes.

Budget and Long-Term Costs

While the upfront cost of a wood stove is a significant consideration, it's equally important to factor in long-term expenses, such as fuel, maintenance, and repairs. Here's how to budget effectively:

1. **Initial Cost**: Wood stoves range from $1,000 to $5,000 or more, depending on size, material, and features.
2. **Fuel Costs**: Seasoned hardwoods are the best choice for efficiency and heat output. Prices vary by region but are generally lower than fossil fuels.
3. **Maintenance**: Regular cleaning, chimney inspections, and potential part replacements (e.g., gaskets or catalytic combustors) should be included in your budget.
4. **Upgrades**: Consider whether additional components, such as air purifiers or heat exchangers, might enhance your stove's performance.

Special Features to Consider

Modern wood stoves offer a range of features that enhance usability and performance. Depending on your needs, you may want to prioritize:

1. **Thermostatic Controls**: Allow precise temperature

regulation for consistent heating.
2. **Ash Pan**: Simplifies ash removal and reduces mess.
3. **Blowers and Fans**: Improve heat distribution, especially in larger spaces.
4. **Cooking Surfaces**: Some stoves include flat tops or oven compartments for dual-purpose functionality.
5. **Smart Technology**: Advanced models integrate with smart home systems for remote monitoring and control.

Evaluating Your Options

Before making a final decision, visit local showrooms or attend home expos to see different models in action. Speaking with experts and reading customer reviews can also provide valuable insights. When possible, test how the stove operates—from loading wood to adjusting controls—to ensure it meets your expectations.

Selecting the right wood stove is about more than just heating your home. It's about creating a space that reflects your values, enhances your comfort, and supports your lifestyle. With careful consideration and informed decision-making, your wood stove can become one of the most rewarding investments you make for your home.

Chapter 4: Proper Installation for Safety and Efficiency

Installing a wood stove is not simply a matter of placing it in a room and lighting a fire. It requires careful planning, meticulous execution, and adherence to safety standards to ensure that the stove operates efficiently and poses no risks to your home or family. A well-installed wood stove not only maximizes heat output but also minimizes emissions, extends the lifespan of the appliance, and ensures compliance with local building codes. This chapter will guide you through every aspect of proper wood stove installation, from selecting the ideal location to understanding chimney dynamics and addressing common pitfalls.

The Importance of Proper Installation

Improperly installed wood stoves can lead to a host of issues, including:

- Increased risk of fire due to proximity to combustible materials.
- Inefficient burning, resulting in excessive smoke, creosote buildup, and higher emissions.
- Carbon monoxide leaks that pose a serious health hazard.
- Voided warranties or non-compliance with insurance and building regulations.

By prioritizing proper installation, you safeguard your home, enhance the stove's performance, and create a more enjoyable heating experience.

Step 1: Selecting the Ideal Location

The placement of your wood stove significantly impacts its efficiency and safety. Consider these factors when choosing a location:

1. **Centralized Position**:
    - Place the stove in a central location to allow heat to distribute evenly throughout the home.
    - Avoid positioning it in isolated corners or rooms with poor airflow, as this limits heat circulation.
2. **Proximity to Chimney**:

- Install the stove near an existing chimney if available, reducing the need for extensive ductwork or modifications.
- Ensure the chimney is in good condition and meets modern safety standards.

3. **Clearances**:
    - Maintain safe distances from walls, furniture, and other combustible materials. Most manufacturers specify minimum clearance requirements, typically ranging from 12 to 36 inches.
    - Use non-combustible wall shields to reduce required clearances if space is limited.

4. **Flooring Considerations**:
    - Install the stove on a fire-resistant surface, such as a concrete slab, tile, or a dedicated hearth pad.
    - Ensure the flooring extends at least 18 inches beyond the front and sides of the stove to catch stray embers or sparks.

Step 2: Choosing the Right Venting System

The venting system is the backbone of a wood stove installation. It removes smoke, gases, and particulates from the stove while supplying fresh air for combustion. A properly designed venting system ensures safety and efficiency.

1. **Understanding Chimney Dynamics**:
   - **Draft**: The chimney draft pulls smoke and gases out of the stove. An efficient draft depends on the height, diameter, and insulation of the chimney.
   - **Height**: The chimney should extend at least three feet above the roofline and two feet higher than any structure within a 10-foot radius to maintain a strong draft.
   - **Insulation**: Insulated chimneys retain heat, which improves draft and reduces creosote buildup.
2. **Types of Chimneys**:
   - **Masonry Chimneys**: Traditional and durable, but may require relining to meet modern safety standards.
   - **Factory-Built Chimneys**: Prefabricated metal chimneys designed for specific stoves, offering easy installation and high efficiency.
3. **Chimney Liners**:
   - Use stainless steel liners to improve draft, reduce creosote accumulation, and protect masonry chimneys from heat damage.

- Ensure the liner matches the stove's flue diameter for optimal performance.

4. **Flue Pipes**:
    - Connect the stove to the chimney using stovepipes made of durable, heat-resistant material.
    - Maintain proper clearances from ceilings and walls, and use double-walled pipes if necessary to reduce heat transfer.

Step 3: Ensuring Adequate Ventilation

Ventilation plays a crucial role in maintaining combustion efficiency and indoor air quality. Poor ventilation can result in incomplete combustion, leading to excessive smoke and carbon monoxide production.

1. **Air Supply**:
    - Ensure the stove has access to adequate fresh air. In airtight homes, consider installing an external air supply vent to feed the fire.
    - Avoid using exhaust fans in the same room as the stove, as they can create negative pressure and disrupt the draft.

2. **Room Size**:
    - Verify that the room size aligns with the stove's specifications. Small, enclosed spaces may require additional ventilation to prevent oxygen depletion.
3. **Carbon Monoxide Detectors**:
    - Install carbon monoxide detectors near the stove and in sleeping areas to monitor air quality and provide early warnings of dangerous gas levels.

Step 4: Following Manufacturer Instructions

Every wood stove is unique, with specific installation requirements outlined by the manufacturer. These guidelines take precedence over general recommendations and ensure compliance with safety standards. Pay close attention to:

1. **Clearance Requirements**:
    - Adhere to specified distances between the stove and combustible materials.
    - Use approved heat shields or barriers if space constraints necessitate reduced clearances.
2. **Flue Diameter**:
    - Match the flue diameter to the stove's

specifications to maintain proper draft and airflow.

3. **Chimney Connections**:
   - Use only manufacturer-approved connectors and components to ensure compatibility and safety.

Step 5: Professional Installation vs. DIY

While some homeowners may be tempted to install a wood stove themselves, professional installation is often the safer and more efficient option. Here's why:

1. **Knowledge of Building Codes**:
   - Certified installers are familiar with local building codes and can ensure compliance.
2. **Expertise**:
   - Professionals have the experience and tools to handle complex installations, reducing the risk of errors.
3. **Warranty Protection**:
   - Many manufacturers require professional installation to maintain warranty coverage.

4. **Safety Inspections**:
    - Installers conduct safety checks to confirm that all components are functioning correctly.

Step 6: Post-Installation Safety Checks

After the stove is installed, conduct a series of safety checks before lighting the first fire:

1. **Inspect Clearances**:
    - Verify that all clearances meet manufacturer guidelines and local codes.
2. **Check Connections**:
    - Ensure all stovepipe and chimney connections are secure and properly sealed.
3. **Test the Draft**:
    - Light a small piece of newspaper in the stove to confirm that the draft is functioning correctly.
4. **Inspect Carbon Monoxide Detectors**:
    - Confirm that detectors are operational and placed in appropriate locations.

Step 7: Maintenance and Upkeep

Proper installation is just the beginning. Regular maintenance is essential to ensure the stove continues to operate safely and efficiently. Key maintenance tasks include:

1. **Chimney Cleaning**:
   - Schedule annual chimney cleanings to remove creosote buildup and prevent blockages.
2. **Stove Inspection**:
   - Inspect the stove for cracks, warped components, or other signs of wear.
3. **Gasket Replacement**:
   - Replace door gaskets as needed to maintain an airtight seal.
4. **Ash Removal**:
   - Empty the ash pan regularly to maintain optimal airflow and prevent ash buildup.

Common Installation Pitfalls to Avoid

Even well-intentioned installations can go awry. Avoid these common mistakes:

1. **Improper Clearances**:
    - Ignoring clearance requirements increases the risk of fire.
2. **Oversized Stoves**:
    - Choosing a stove that's too large for the space can lead to overheating and inefficient burning.
3. **Poor Venting**:
    - Inadequate venting reduces draft efficiency and increases smoke production.
4. **Neglecting Local Codes**:
    - Failure to comply with building codes can result in fines, insurance issues, or safety hazards.

By following these detailed guidelines and prioritizing safety and efficiency, you can ensure that your wood stove becomes a reliable and enjoyable feature of your home.

Chapter 5: Fueling the Fire: Choosing the Right Wood

The heart of any wood stove is its fuel, and in this case, that fuel is wood. While it might seem straightforward—wood is wood, right?—the type, quality, and preparation of the wood you burn can significantly impact the performance, efficiency, and safety of your stove. Choosing the right wood isn't just about finding something that burns; it's about understanding the characteristics of different types of wood, how to properly prepare it, and the best practices for storage and usage. In this chapter, we'll explore the science and art of fueling your fire with the right wood, equipping you with the knowledge to get the most out of your stove.

The Basics of Firewood

All wood is made of organic material that, when burned, undergoes a chemical reaction releasing heat, light, and gases. However, not all wood is created equal. The properties of firewood can vary greatly depending on the species, density, moisture content, and preparation.

1. **Softwoods vs. Hardwoods**:
   - **Softwoods**: Derived from coniferous trees like pine, spruce, and fir. These woods are less dense and ignite quickly, making them ideal for kindling. However, they burn faster and produce more resin, which can lead to creosote buildup in your chimney.
   - **Hardwoods**: Sourced from deciduous trees such as oak, maple, and birch. Hardwoods are denser, burn longer, and produce more heat per log. They're the preferred choice for sustained heating.
2. **Energy Content (BTUs)**:
   - Firewood is often measured in terms of its energy content, typically in British Thermal Units (BTUs). Hardwoods like oak and hickory have higher BTU ratings compared to softwoods, meaning they provide more heat per volume.
3. **Moisture Content**:
   - The moisture content of wood is one of the most critical factors in its performance. Freshly cut, or "green," wood can have moisture levels above 50%, which results in poor combustion, excessive smoke, and reduced heat output. Properly

seasoned wood should have a moisture content of 20% or less.

The Best Woods for Your Stove

Not all woods are equally suitable for burning. Here's a breakdown of some of the best options:

1. **Oak**:
    - **Characteristics**: Dense, slow-burning, and produces a steady heat.
    - **Pros**: High heat output, long burn time, minimal smoke.
    - **Cons**: Takes longer to season properly (at least 1-2 years).
2. **Maple**:
    - **Characteristics**: Hard, burns cleanly, and provides a moderate-to-high heat output.
    - **Pros**: Easy to split and season, produces little ash.
    - **Cons**: Slightly less heat output compared to oak.
3. **Hickory**:
    - **Characteristics**: Among the highest BTU ratings, burns long and hot.

- **Pros**: Exceptional heat and long burn time.
- **Cons**: Can be challenging to split.

4. **Birch**:
    - **Characteristics**: Burns quickly but produces high heat.
    - **Pros**: Easy to light, ideal for mixing with slower-burning woods.
    - **Cons**: Burns faster, requiring more frequent refueling.

5. **Ash**:
    - **Characteristics**: Burns well even when green, though better when seasoned.
    - **Pros**: Easy to split, consistent heat output.
    - **Cons**: Moderate burn time compared to oak or hickory.

6. **Cherry**:
    - **Characteristics**: Produces a pleasant aroma and moderate heat.
    - **Pros**: Aesthetic flames, low creosote production.
    - **Cons**: Slightly lower heat output.

Woods to Avoid

Certain types of wood are unsuitable for burning due to their characteristics or environmental impact:

1. **Green Wood**:
    - High moisture content results in poor combustion, excessive smoke, and creosote buildup.
2. **Treated or Painted Wood**:
    - Releases toxic chemicals when burned, posing health risks and damaging the stove.
3. **Driftwood**:
    - Contains salt, which can corrode metal components and produce harmful emissions.
4. **Softwoods with High Resin Content**:
    - Such as pine, which can produce creosote quickly if burned in large quantities.

The Science of Seasoning Firewood

Seasoning firewood is the process of reducing its moisture content to improve its burn quality. Properly seasoned wood is more efficient, burns cleaner, and produces less creosote. Here's how to do it effectively:

1. **Splitting**:
    - Split logs into smaller pieces to expose more surface area, speeding up the drying process.
2. **Stacking**:
    - Stack wood off the ground using pallets or logs to prevent moisture absorption from the soil.
    - Leave gaps between logs for airflow, which aids drying.
    - Stack in a crisscross pattern to maximize ventilation.
3. **Covering**:
    - Protect the stack from rain and snow using a tarp, but leave the sides open for airflow.
4. **Time**:
    - Most hardwoods require 6-12 months to season properly, while denser woods like oak may need up to 2 years.
5. **Moisture Meters**:
    - Invest in a moisture meter to check the wood's moisture content. Aim for 20% or less before burning.

Storage Best Practices

Proper storage ensures that your firewood remains dry and ready to burn:

1. **Location**:
    - Store wood in a well-ventilated area, away from the house to prevent pest infestations.
    - Avoid storing wood indoors for extended periods, as this can introduce insects and mold.
2. **Firewood Sheds**:
    - Build or purchase a firewood shed that offers protection from the elements while allowing airflow.
3. **Rotation**:
    - Use older wood first to prevent it from rotting or becoming overly dry.

Using Wood Efficiently

Burning wood efficiently maximizes heat output and minimizes waste. Here's how to make the most of your fuel:

1. **Building the Fire**:
   - Start with kindling and small pieces of wood to establish a strong base.
   - Gradually add larger logs once the fire is burning steadily.
2. **Avoid Overloading**:
   - Overfilling the stove can smother the fire and lead to incomplete combustion.
3. **Burning in Cycles**:
   - Allow the fire to burn down to hot coals before adding more wood. This ensures a clean and efficient burn.
4. **Air Control**:
   - Adjust the stove's air vents to regulate oxygen flow. Too little air can cause smoldering, while too much can waste heat.

Sustainability and Ethical Wood Sourcing

Using firewood responsibly is not just about personal efficiency; it's also about environmental stewardship:

1. **Local Sourcing**:

- Purchase wood from local suppliers to reduce transportation emissions and support sustainable forestry practices.

2. **Replanting**:
    - Choose suppliers who engage in reforestation efforts to replenish harvested trees.
3. **Alternative Fuels**:
    - Consider using compressed wood logs or pellets made from sawdust and wood waste for a more sustainable option.

The Role of Wood in Stove Efficiency

The quality of your firewood directly impacts your stove's performance:

1. **Heat Output**:
    - Dry, dense hardwoods produce the most consistent heat.
2. **Creosote Prevention**:
    - Burning seasoned wood reduces creosote buildup, minimizing chimney maintenance and fire risk.

3. **Smoke and Emissions**:
    - Properly prepared wood burns cleaner, improving indoor air quality and reducing environmental impact.

Choosing the right wood is as much an art as it is a science. With the knowledge shared in this chapter, you'll be equipped to fuel your fire effectively, enhancing the efficiency of your stove and the comfort of your home.

Chapter 6: Maintaining Your Wood Stove for

Longevity and Health

A wood stove is more than just a heating appliance—it's a long-term investment that requires proper care to function efficiently and safely. Regular maintenance not only extends the life of your stove but also ensures it operates in an environmentally friendly and health-conscious manner. In this chapter, we'll delve into every aspect of wood stove maintenance, from routine cleaning to advanced troubleshooting, equipping you with the knowledge to keep your stove in peak condition.

Why Maintenance Matters

Neglecting wood stove maintenance can lead to several issues, including:

1. **Inefficient Heating**: A poorly maintained stove burns wood less efficiently, leading to wasted fuel and reduced heat output.
2. **Increased Emissions**: Creosote buildup and incomplete combustion result in higher emissions, contributing to indoor air pollution and environmental harm.
3. **Health Risks**: Dirty stoves and chimneys can produce carbon monoxide, a colorless and odorless gas that

poses severe health risks.
4. **Safety Hazards**: Creosote and ash buildup increase the likelihood of chimney fires.
5. **Reduced Longevity**: Regular maintenance prevents wear and tear, prolonging the life of your stove.

Daily Maintenance: Keeping Your Stove Running Smoothly

Daily care ensures your wood stove operates at its best during the heating season. These simple tasks take only a few minutes but make a significant difference:

1. **Emptying the Ash Pan**:
    - Ash buildup can restrict airflow, reducing combustion efficiency.
    - Wait until the stove is completely cool before removing ashes.
    - Leave a thin layer of ash on the bottom to insulate the firebox and help maintain heat.
2. **Inspecting the Firebox**:
    - Check for cracks or warping in the firebox that could compromise safety or efficiency.
    - Remove any debris or unburned wood remnants.

3. **Checking Air Vents**:
    - Ensure air vents are free of obstructions to maintain proper oxygen flow.
    - Adjust vents to optimize combustion and heat output.

Weekly Maintenance: Keeping Efficiency High

Set aside time each week to perform slightly more in-depth maintenance tasks:

1. **Cleaning the Glass**:
    - Use a soft cloth or newspaper dampened with a vinegar-and-water solution to clean the stove's glass door.
    - Avoid abrasive cleaners that could scratch the surface.
    - Regular cleaning prevents soot buildup and maintains a clear view of the flames.
2. **Checking Gaskets**:
    - Inspect the door gasket for signs of wear, such as fraying or compression.
    - Replace damaged gaskets to maintain an airtight

seal and improve efficiency.
3. **Removing Minor Creosote Deposits**:
    - Use a creosote removal log or brush to tackle early-stage buildup in the chimney and stovepipe.
    - Regular removal prevents larger, more hazardous deposits from forming.

Monthly Maintenance: Ensuring Safe Operation

Monthly maintenance helps catch potential problems before they escalate:

1. **Inspecting the Chimney**:
    - Use a flashlight to look for creosote buildup, blockages, or damage inside the chimney.
    - Ensure the chimney cap is secure and free of debris.
2. **Checking the Flue**:
    - Confirm that the flue opens and closes smoothly.
    - Remove any obstructions, such as bird nests or leaves, that could impede airflow.
3. **Testing Carbon Monoxide Detectors**:
    - Replace batteries in carbon monoxide detectors

and test their functionality.
- Position detectors near the stove and in sleeping areas for maximum safety.

Seasonal Maintenance: Preparing for Peak Performance

At the start and end of each heating season, perform a thorough inspection and deep cleaning to ensure your stove is ready for consistent use:

1. **Deep Cleaning the Stove**:
    - Remove all ashes, soot, and debris from the firebox, ash pan, and stovepipe.
    - Clean internal components, such as baffles and secondary combustion chambers.
2. **Inspecting Components**:
    - Check all moving parts, such as door hinges and air controls, for wear or damage.
    - Lubricate hinges and handles as needed.
3. **Professional Chimney Inspection**:
    - Hire a certified chimney sweep to clean and inspect the chimney thoroughly.
    - Ensure the chimney lining is intact and free of

cracks or deterioration.
4. **Repainting or Touching Up**:
    - Use high-temperature paint to touch up any scratches or areas of rust on the stove's surface.
    - This prevents corrosion and maintains the stove's appearance.

Common Maintenance Challenges and Solutions

Even with regular care, you may encounter issues that require attention. Here's how to address common problems:

1. **Creosote Buildup**:
    - **Cause**: Burning wet or unseasoned wood, poor draft, or low burn temperatures.
    - **Solution**: Use seasoned hardwood, maintain a hot fire, and clean the chimney regularly.
2. **Smoke Backdraft**:
    - **Cause**: Negative air pressure, blocked chimney, or improper venting.
    - **Solution**: Ensure proper chimney height, clear blockages, and provide adequate ventilation.
3. **Difficulty Lighting the Fire**:

- **Cause**: Damp wood, insufficient kindling, or poor airflow.
- **Solution**: Use dry, seasoned wood and arrange kindling to promote airflow.
4. **Warped or Cracked Components**:
    - **Cause**: Overfiring the stove or using improper fuels.
    - **Solution**: Replace damaged parts and avoid overloading the stove.

Health and Environmental Considerations

Maintaining your wood stove isn't just about performance; it's also about safeguarding your health and minimizing environmental impact:

1. **Reducing Emissions**:
    - Burn only dry, seasoned wood to reduce smoke and particulates.
    - Avoid burning trash, treated wood, or other materials that release toxic fumes.

2. **Improving Indoor Air Quality**:

- Use air purifiers with HEPA filters to capture fine particulates.
- Ensure proper ventilation to maintain fresh indoor air.

3. **Preventing Carbon Monoxide Poisoning**:
    - Install carbon monoxide detectors and test them regularly.
    - Never let the stove smolder for long periods, as this can produce dangerous gases.

Upgrading and Modernizing Your Stove

Over time, you may consider upgrading your wood stove to incorporate modern technologies that improve efficiency and safety:

1. **EPA-Certified Stoves**:
    - Modern stoves produce fewer emissions and are up to 50% more efficient than older models.

2. **Catalytic Converters**:
    - Adding a catalytic converter can improve

combustion efficiency and reduce creosote.

3. **Smart Features**:
    - Some stoves now include digital controls, temperature sensors, and remote monitoring capabilities.

Creating a Maintenance Schedule

Consistency is key to effective maintenance. Create a schedule that outlines daily, weekly, monthly, and seasonal tasks to ensure nothing is overlooked. Use reminders or apps to track your progress and stay on top of essential upkeep.

By following these comprehensive maintenance practices, you can enjoy the warmth and ambiance of your wood stove while ensuring it remains a safe and efficient addition to your home.

Chapter 7: Understanding Indoor Air Pollution

When most people think about pollution, their minds wander to crowded cities, car exhaust fumes, and industrial smokestacks. But often overlooked is the air inside our homes—a space where we spend the majority of our time. Indoor air pollution is a silent but potent threat, capable of impacting health and well-being in profound ways. Understanding its sources, effects, and mitigation strategies is critical, particularly for households with wood stoves. This chapter delves deep into the science of indoor air pollution, its implications, and actionable solutions to create a healthier living environment.

What is Indoor Air Pollution?

Indoor air pollution refers to the presence of harmful pollutants within enclosed spaces. Unlike outdoor air, indoor air is often trapped and recirculated, allowing contaminants to accumulate over time. This makes indoor environments particularly susceptible to pollution-related issues.

Key indoor pollutants include:

1. **Particulate Matter (PM):**
    - Fine particles (PM2.5) and larger particles (PM10) that can be inhaled into the lungs.
    - Sources: Smoke from wood stoves, cooking, and dust.
2. **Volatile Organic Compounds (VOCs):**
    - Organic chemicals that vaporize at room temperature.
    - Sources: Paints, cleaning products, furniture, and even burning wood.
3. **Carbon Monoxide (CO):**
    - A colorless, odorless gas that can be deadly in high concentrations.
    - Sources: Incomplete combustion from wood stoves, gas stoves, and heaters.
4. **Nitrogen Dioxide (NO_2):**
    - A reactive gas formed during combustion processes.
    - Sources: Wood stoves, gas appliances, and cigarette smoke.
5. **Biological Pollutants:**
    - Includes mold spores, bacteria, pet dander, and pollen.

- Sources: Damp environments, pets, and poor ventilation.

How Indoor Air Pollution Impacts Health

Indoor air pollution affects everyone, but certain groups—such as children, the elderly, and individuals with pre-existing health conditions—are particularly vulnerable. The health effects range from minor irritations to severe chronic illnesses.

1. **Short-Term Effects:**
    - Irritation of the eyes, nose, and throat.
    - Headaches, dizziness, and fatigue.
    - Respiratory symptoms such as coughing and wheezing.
2. **Long-Term Effects:**
    - Chronic respiratory diseases, including asthma and bronchitis.
    - Cardiovascular issues, such as heart disease and hypertension.
    - Developmental and cognitive delays in children.
    - Increased risk of cancer from prolonged exposure to certain pollutants.

3. **Silent Threats:**
   - Carbon monoxide poisoning can occur without warning, leading to confusion, unconsciousness, or even death if not detected.
   - Long-term exposure to fine particulates (PM2.5) can reduce lung capacity and accelerate aging-related health declines.

Wood Stoves and Indoor Air Pollution

While wood stoves offer warmth and charm, they can be significant contributors to indoor air pollution if not used and maintained correctly. Here's how wood stoves impact air quality:

1. **Particulate Matter:**
   - Inefficient burning produces smoke laden with fine particles that can linger in the air.
2. **Creosote Emissions:**
   - Incomplete combustion generates creosote, which can off-gas harmful chemicals.
3. **Carbon Monoxide:**
   - Poor ventilation or faulty stove components can result in CO leakage.

4. **VOCs:**
   - Burning certain types of wood or debris releases volatile organic compounds.

Mitigating Pollution from Wood Stoves

1. **Use Seasoned Wood:**
   - Always burn wood with a moisture content of 20% or less to reduce smoke and emissions.
2. **Ensure Proper Ventilation:**
   - Install an appropriate chimney system and ensure it's well-maintained.
3. **Upgrade to EPA-Certified Stoves:**
   - These models burn more efficiently and emit fewer pollutants.
4. **Monitor Air Quality:**
   - Use indoor air quality monitors to detect elevated levels of particulates or gases.

Common Sources of Indoor Air Pollution Beyond Wood Stoves

1. **Cooking and Kitchen Activities:**
   - Gas stoves and ovens release nitrogen dioxide and carbon monoxide.

- Frying and grilling produce airborne grease and particulates.

2. **Household Products:**
   - Cleaning sprays, air fresheners, and pesticides often contain harmful VOCs.
3. **Building Materials:**
   - New furniture, carpets, and paints can off-gas formaldehyde and other chemicals.
4. **Outdoor Pollutants:**
   - Pollutants such as pollen, dust, and vehicle emissions can infiltrate indoor spaces.

Strategies for Reducing Indoor Air Pollution

1. **Improve Ventilation:**
   - Use exhaust fans in kitchens and bathrooms.
   - Open windows when weather permits to exchange stale air with fresh outdoor air.
   - Install mechanical ventilation systems, such as HRVs (Heat Recovery Ventilators).
2. **Invest in Air Purifiers:**
   - Choose models with HEPA filters to capture fine particulates.

- Opt for units with activated carbon filters to remove VOCs and odors.

3. **Control Humidity Levels:**
    - Maintain indoor humidity between 30-50% to prevent mold growth.
    - Use dehumidifiers or air conditioners in damp areas.

4. **Adopt Healthier Habits:**
    - Avoid smoking indoors.
    - Limit the use of candles and incense.
    - Use non-toxic cleaning products and natural air fresheners.

5. **Regular Maintenance:**
    - Clean air ducts and replace HVAC filters regularly.
    - Inspect and maintain appliances that involve combustion, such as wood stoves and gas heaters.

Emerging Technologies in Indoor Air Quality Management

Advancements in technology are offering new ways to monitor and improve indoor air quality:

1. **Smart Air Quality Monitors:**

- Devices that provide real-time data on pollutant levels, humidity, and temperature.
2. **Advanced Filtration Systems:**
    - Electrostatic filters and UV-C light systems that neutralize bacteria and viruses.
3. **Smart Ventilation Systems:**
    - Integrate with home automation to optimize airflow and filtration.
4. **Natural Air Purifiers:**
    - Innovations like biofilters using plants and activated charcoal to clean air organically.

The Role of Awareness and Education

Understanding indoor air pollution is the first step toward mitigating its effects. Education empowers individuals to make informed decisions about:

- The materials and appliances they bring into their homes.
- The habits they adopt to maintain a clean living environment.
- The investments they make in air quality improvement tools.

Promoting awareness within communities can have a ripple effect, inspiring widespread adoption of healthier practices.

By grasping the complexities of indoor air pollution and implementing proactive measures, you can create a home environment that is not only warm and inviting but also safe and healthy for everyone who lives there.

Chapter 8: The Role of Portable Nebulizers in

Respiratory Health

Respiratory health is one of the cornerstones of overall well-being. For individuals living with conditions like asthma, chronic obstructive pulmonary disease (COPD), or seasonal allergies, maintaining clear airways can be a daily challenge. The air we breathe, both indoors and outdoors, plays a critical role in our respiratory function, and when air quality is compromised—whether by pollution, allergens, or wood stove emissions—it becomes even more essential to have tools that support respiratory health. Among these tools, portable nebulizers have emerged as an invaluable resource for managing respiratory conditions and improving quality of life.

In this chapter, we'll explore the science behind nebulizers, their role in respiratory health, and how they can be integrated into a wellness routine. We'll also delve into the practical benefits of portable nebulizers and why they're particularly suited for today's fast-paced and mobile lifestyles.

What is a Nebulizer?

A nebulizer is a medical device designed to deliver medication directly to the lungs in the form of a fine mist. This method of drug delivery ensures rapid absorption, making nebulizers an effective treatment option for a range of respiratory conditions. Nebulizers are commonly used to administer bronchodilators, corticosteroids, and other medications that alleviate symptoms like wheezing, shortness of breath, and chest tightness.

1. **How Nebulizers Work:**
    - The device uses compressed air, ultrasonic vibrations, or a mesh system to transform liquid medication into an aerosolized mist.
    - This mist is then inhaled through a mask or mouthpiece, allowing the medication to reach the lower respiratory tract.
2. **Types of Nebulizers:**
    - **Jet Nebulizers:** Use compressed air to generate the mist. They're reliable but often bulky.
    - **Ultrasonic Nebulizers:** Employ high-frequency sound waves to aerosolize medication. They're quieter and faster but not suitable for all types of drugs.
    - **Mesh Nebulizers:** Utilize a vibrating mesh to produce a fine mist. These are compact, portable,

and highly efficient, making them ideal for personal use.

The Growing Popularity of Portable Nebulizers

Portable nebulizers have revolutionized respiratory care by offering convenience and flexibility. Unlike traditional models, which are often large and require an external power source, portable nebulizers are lightweight, battery-operated, and designed for on-the-go use. This makes them particularly valuable for individuals who travel frequently or have active lifestyles.

1. **Key Features of Portable Nebulizers:**
    - Compact design for easy storage and transportation.
    - Rechargeable batteries or USB power options.
    - Quiet operation, ideal for discreet use in public settings.
    - Rapid treatment times, often completing a session in 5-10 minutes.
2. **Who Benefits Most from Portable Nebulizers?**
    - Individuals with chronic respiratory conditions like asthma or COPD.

- Parents of young children who require frequent treatments.
- Athletes and outdoor enthusiasts exposed to environmental triggers.
- Elderly individuals who prefer a user-friendly and low-maintenance device.

Conditions Treated with Nebulizers

Nebulizers are a versatile tool in respiratory care, addressing a wide range of conditions. Here are some of the most common:

1. **Asthma:**
   - Nebulizers deliver bronchodilators and corticosteroids to relieve airway inflammation and prevent asthma attacks.
   - They are particularly useful during severe episodes when rapid medication delivery is critical.
2. **Chronic Obstructive Pulmonary Disease (COPD):**
   - For individuals with COPD, nebulizers help manage chronic symptoms like persistent cough and mucus buildup.
   - Long-acting bronchodilators can improve lung

function and overall quality of life.
3. **Seasonal Allergies:**
   - Nebulizers can deliver antihistamines or corticosteroids to reduce inflammation caused by allergens like pollen, mold, or pet dander.
4. **Cystic Fibrosis:**
   - Nebulized saline solutions and antibiotics help thin mucus and prevent respiratory infections in individuals with cystic fibrosis.
5. **Acute Respiratory Infections:**
   - During illnesses like bronchitis or pneumonia, nebulizers can administer medications to reduce inflammation and improve breathing.
6. **Other Uses:**
   - Some individuals use nebulizers for non-medical treatments, such as inhaling saline solutions to hydrate airways or essential oils for aromatherapy (though this should only be done under professional guidance).

Advantages of Portable Nebulizers

Portable nebulizers offer a host of benefits that make them

indispensable for respiratory health management:

1. **Convenience:**
    - Their compact size and lightweight design make them easy to carry in a bag or pocket.
    - Battery-operated models allow for treatments anywhere, whether at home, work, or while traveling.
2. **Ease of Use:**
    - Modern nebulizers are user-friendly, with simple controls and quick assembly.
    - Suitable for children, elderly individuals, and those with limited dexterity.
3. **Rapid Symptom Relief:**
    - Nebulizers provide immediate relief during respiratory emergencies, such as asthma attacks.
4. **Customizable Treatment:**
    - Compatible with a variety of medications, allowing for tailored treatment plans.
5. **Quiet Operation:**
    - Many portable nebulizers are designed for silent or low-noise operation, ensuring discretion in public settings.

Integrating Portable Nebulizers into Your Routine

For individuals managing chronic respiratory conditions, incorporating a portable nebulizer into daily life can enhance treatment adherence and overall health. Here's how to make the most of your device:

1. **Develop a Treatment Schedule:**
    - Work with your healthcare provider to establish a routine that aligns with your needs and lifestyle.
    - Set reminders to ensure treatments are taken consistently.
2. **Maintain Proper Hygiene:**
    - Clean the nebulizer components after each use to prevent bacterial growth and ensure optimal performance.
    - Replace filters, tubing, and masks as recommended by the manufacturer.
3. **Use as Directed:**
    - Follow dosage and medication guidelines provided by your doctor to avoid under- or over-treatment.
4. **Monitor Symptoms:**
    - Keep track of your respiratory symptoms to identify

triggers and assess treatment effectiveness.
5. **Combine with Other Wellness Practices:**
    - Pair nebulizer treatments with other respiratory health strategies, such as staying hydrated, maintaining good indoor air quality, and practicing breathing exercises.

Nebulizers and Environmental Factors

Environmental conditions can exacerbate respiratory issues, making portable nebulizers even more valuable in mitigating symptoms:

1. **Air Pollution:**
    - Nebulizers provide relief after exposure to pollutants like PM2.5, ozone, and nitrogen dioxide.
2. **Seasonal Changes:**
    - During allergy seasons, nebulizers help manage inflammation caused by pollen and other allergens.
3. **Indoor Air Quality:**
    - For households with wood stoves or other

combustion-based heating, nebulizers can alleviate respiratory irritation from particulate matter and smoke.

4. **Travel-Related Triggers:**
    - Portable nebulizers are a lifeline for individuals traveling to areas with poor air quality or different climates.

Choosing the Right Portable Nebulizer

With various models on the market, selecting the best nebulizer for your needs requires careful consideration:

1. **Size and Portability:**
    - Opt for a lightweight and compact design if you plan to use it while traveling.
2. **Battery Life:**
    - Look for models with long-lasting rechargeable batteries or USB power options.
3. **Noise Level:**
    - Choose a quiet device if you need discretion or plan to use it in shared spaces.
4. **Medication Compatibility:**

- Ensure the nebulizer is suitable for your prescribed medications, especially if they have specific delivery requirements.
5. **Ease of Cleaning:**
    - Select a model with detachable, dishwasher-safe components for easier maintenance.
6. **Cost and Warranty:**
    - Compare prices and check for warranties to ensure long-term reliability.

Future Innovations in Portable Nebulizers

The field of respiratory health is constantly evolving, and nebulizer technology is no exception. Emerging trends and advancements include:

1. **Smart Nebulizers:**
    - Devices equipped with Bluetooth connectivity to track treatment adherence and provide feedback.
2. **Wearable Nebulizers:**
    - Compact, hands-free models that can be worn discreetly for continuous treatment.
3. **Integrated Air Quality Monitors:**

- Nebulizers with built-in sensors to detect pollutants and recommend treatment adjustments.
4. **Biodegradable Components:**
    - Environmentally friendly designs that reduce plastic waste.

By embracing the versatility and convenience of portable nebulizers, individuals can take control of their respiratory health, ensuring that they breathe easier regardless of their environment or circumstances.

Chapter 9: Creating a Healthy Home Environment

Our homes are more than just shelters; they are spaces where we relax, rejuvenate, and spend the majority of our lives. A healthy home environment is essential for physical well-being, mental health, and overall quality of life. Yet, achieving such an environment requires intentional design, regular maintenance, and thoughtful choices about everything from air quality to materials used in construction and decoration. This chapter explores the key elements of a healthy home environment and offers practical strategies for creating a space that promotes health and comfort for all its occupants.

The Foundations of a Healthy Home

A healthy home environment is built on three primary principles:

1. **Clean Air:** The air inside your home should be free from pollutants, allergens, and toxins. Indoor air quality is often worse than outdoor air, making it critical to address sources of contamination and ensure proper ventilation.
2. **Safe Materials:** From paint to furniture, the materials used in your home can emit volatile organic compounds (VOCs) and other harmful substances. Opting for non-toxic, sustainable options minimizes exposure to these hazards.
3. **Comfortable Design:** A well-designed home supports mental and physical health by fostering relaxation,

reducing stress, and enhancing functionality.

Optimizing Indoor Air Quality

The air we breathe inside our homes has a profound impact on our health. Poor indoor air quality can cause respiratory problems, exacerbate allergies, and even contribute to chronic illnesses. To optimize air quality, consider the following strategies:

1. **Ventilation:**
    - Open windows regularly to allow fresh air to circulate and reduce the concentration of indoor pollutants.
    - Install mechanical ventilation systems, such as heat recovery ventilators (HRVs) or energy recovery ventilators (ERVs), to exchange stale indoor air with fresh outdoor air efficiently.

2. **Air Purification:**
    - Use air purifiers equipped with HEPA filters to capture fine particulates, allergens, and pet dander.
    - Choose models with activated carbon filters to

absorb VOCs and odors.
3. **Humidity Control:**
    - Maintain indoor humidity levels between 30-50% to prevent mold growth and reduce dust mites.
    - Use dehumidifiers in damp areas and humidifiers in dry climates to achieve balanced humidity levels.
4. **Eliminating Pollutants:**
    - Avoid smoking indoors and limit the use of candles or incense.
    - Use exhaust fans in kitchens and bathrooms to remove cooking fumes and excess moisture.
    - Regularly clean and vacuum using a machine with a HEPA filter to reduce dust and allergens.

Choosing Non-Toxic and Sustainable Materials

The materials used in your home can significantly impact your health. Many conventional products release harmful chemicals over time, contributing to indoor air pollution. Opt for safer, more sustainable options wherever possible:

1. **Paints and Finishes:**
    - Choose low-VOC or zero-VOC paints and finishes

to reduce the release of toxic fumes.
    - Look for products certified by reputable organizations, such as GREENGUARD or Green Seal.
2. **Flooring:**
    - Use natural materials like hardwood, bamboo, or cork, which are durable and low in VOC emissions.
    - Avoid vinyl flooring, which can release phthalates and other harmful chemicals.
3. **Furniture:**
    - Select solid wood furniture instead of particleboard or MDF, which often contain formaldehyde-based adhesives.
    - Opt for upholstery made from natural fibers like cotton, wool, or hemp.
4. **Cleaning Products:**
    - Switch to non-toxic, biodegradable cleaning products that are free from harsh chemicals.
    - Make DIY cleaning solutions using ingredients like vinegar, baking soda, and essential oils.

Managing Lighting for Health and Productivity

Lighting plays a critical role in regulating circadian rhythms, supporting mental health, and creating a comfortable living environment. By optimizing lighting, you can enhance both mood and productivity:

1. **Natural Light:**
    - Maximize exposure to natural light by keeping windows clean and using light-colored window treatments.
    - Arrange furniture to take advantage of daylight for activities like reading or working.
2. **Artificial Lighting:**
    - Use LED bulbs with adjustable brightness and color temperature to mimic natural light.
    - Install dimmer switches to customize lighting levels for different times of the day.
    - Choose warm-toned lighting for relaxation areas and cooler tones for workspaces.
3. **Avoiding Blue Light:**
    - Limit exposure to blue light from screens and LED lighting in the evening, as it can disrupt sleep patterns.
    - Use blue light-blocking glasses or screen filters if evening screen use is unavoidable.

Creating a Restorative Sleep Environment

Sleep is a cornerstone of health, and your bedroom should be a sanctuary designed to promote restorative rest. Here are some tips for creating an optimal sleep environment:

1. **Comfortable Bedding:**
    - Invest in a high-quality mattress and pillows that provide proper support and comfort.
    - Use natural, breathable fabrics for sheets and blankets to regulate body temperature.
2. **Darkness and Quiet:**
    - Install blackout curtains or shades to block out external light.
    - Use white noise machines or earplugs to minimize disruptive sounds.
3. **Temperature Control:**
    - Keep the bedroom cool, ideally between 60-67°F (15-19°C), for optimal sleep conditions.
4. **Electronics-Free Zone:**
    - Avoid keeping televisions, smartphones, or other electronic devices in the bedroom to reduce distractions and blue light exposure.

Water Quality and Hydration

Clean, safe water is essential for both hydration and household activities. Improving water quality ensures better health and enhances the functionality of your home:

1. **Water Filtration:**
    - Install a whole-house filtration system or individual filters for drinking water taps to remove contaminants like chlorine, lead, and pesticides.
    - Use refrigerator water filters or portable filtration pitchers as a more affordable alternative.
2. **Hydration Stations:**
    - Place water dispensers or bottles in accessible locations to encourage regular hydration.
    - Add natural flavors like lemon, cucumber, or mint to make drinking water more appealing.

Incorporating Biophilic Design

Biophilic design integrates elements of nature into indoor spaces, fostering a connection to the outdoors and improving mental health:

1. **Indoor Plants:**
   - Add houseplants like snake plants, peace lilies, or pothos to purify air and bring greenery indoors.
   - Use hanging planters or vertical gardens to maximize space.
2. **Natural Materials:**
   - Incorporate wood, stone, and other natural materials into furniture, decor, and finishes.
3. **Nature-Inspired Decor:**
   - Use artwork, textiles, and color schemes that evoke natural landscapes.
4. **Outdoor Views:**
   - Position seating areas near windows with views of gardens or greenery.

Minimizing Electromagnetic Radiation

While the science around electromagnetic radiation (EMR) and its health effects is still evolving, minimizing exposure is a precautionary measure many households choose to take:

1. **Router Placement:**
   - Keep Wi-Fi routers away from bedrooms and high-

traffic areas.

2. **Device Usage:**
    - Turn off devices when not in use and avoid keeping them close to the body.
3. **Wired Alternatives:**
    - Use wired connections for internet and peripherals instead of relying solely on wireless devices.

Establishing Healthy Habits

Creating a healthy home environment isn't just about physical changes; it's also about adopting habits that promote well-being:

1. **Decluttering:**
    - Keep spaces organized to reduce stress and create a sense of calm.
    - Donate or recycle items you no longer need.
2. **Regular Cleaning:**
    - Develop a cleaning schedule to maintain hygiene and prevent allergens from accumulating.
3. **Mindful Living:**
    - Incorporate mindfulness practices, such as meditation or yoga, into your daily routine to

enhance mental health.

By focusing on these elements, you can transform your home into a haven of health and comfort. In the next chapter, we will explore how to prepare for emergencies with a wood stove, ensuring safety and resilience during challenging times.

Chapter 10: Emergency Preparedness with Wood Stoves

In times of crisis, whether caused by natural disasters, power outages, or extreme weather conditions, a wood stove can be more than just a source of warmth—it can become a lifeline. Unlike electric or gas-powered heating systems, a wood stove operates independently of external energy sources, making it an invaluable asset during emergencies. But leveraging the full potential of a wood stove during such times requires preparation, knowledge, and the right set of tools. This chapter will guide you through the essentials of emergency preparedness with wood stoves, helping you ensure your home remains a safe, warm, and functional haven in any situation.

The Importance of Emergency Preparedness

Emergency scenarios can arise unexpectedly, leaving households without access to critical utilities such as electricity, gas, or running water. A wood stove provides:

1. **Reliable Heat:**
    - Unlike electric heaters, wood stoves remain operational during power outages, ensuring that your home stays warm even in sub-zero temperatures.
2. **Cooking Capabilities:**
    - Many wood stoves feature flat tops or built-in ovens that can be used for cooking food or boiling water during emergencies.
3. **Sustainability:**

- Wood is a renewable resource that can often be sourced locally, providing a steady fuel supply even when other resources are scarce.
4. **Light and Comfort:**
    - The flickering glow of a fire can provide light and a sense of security during otherwise stressful times.

Preparing Your Wood Stove for Emergencies

Before an emergency occurs, it's crucial to ensure that your wood stove is in optimal condition. Regular maintenance and advance preparation can make all the difference.

1. **Inspection and Cleaning:**
    - **Annual Inspections:** Schedule a professional inspection of your wood stove and chimney each year to check for cracks, creosote buildup, or other issues.
    - **Chimney Cleaning:** Ensure the chimney is free of blockages, which can impede airflow and increase the risk of smoke or carbon monoxide poisoning.
2. **Stocking Up on Firewood:**
    - **Seasoned Wood:** Keep a well-stocked supply of

seasoned hardwood, such as oak or maple, to ensure efficient burning during an extended emergency.
- **Storage:** Store wood in a dry, well-ventilated area protected from rain and snow. Use tarps or a dedicated wood shed to keep your supply ready for use.

3. **Essential Accessories:**
   - **Fire-Starting Tools:** Have matches, lighters, and fire-starting aids like kindling or fire starter bricks readily available.
   - **Maintenance Tools:** Keep a poker, ash shovel, and brush nearby for tending the fire and cleaning the stove.
   - **Safety Equipment:** Ensure you have a fire extinguisher, heat-resistant gloves, and carbon monoxide detectors installed near the stove.

Using a Wood Stove Safely During Emergencies

Safety is paramount when relying on a wood stove during an emergency. Follow these guidelines to minimize risks:

1. **Proper Ventilation:**
   - Always ensure the flue is open to allow smoke and gases to exit the chimney.
   - Avoid blocking air vents, as proper airflow is essential for combustion and ventilation.
2. **Avoid Overloading:**
   - Resist the temptation to overfill the stove with wood, as this can lead to overheating or an uncontrolled fire.
3. **Supervise the Fire:**
   - Never leave the stove unattended while it's in use, especially with children or pets in the home.
4. **Monitor Carbon Monoxide Levels:**
   - Install battery-operated or plug-in carbon monoxide detectors with battery backup near the stove and in sleeping areas.
   - Test alarms regularly to ensure they are functioning properly.
5. **Prevent Chimney Fires:**
   - Burn only seasoned hardwood to minimize creosote buildup.
   - Avoid burning trash, cardboard, or wet wood, which can produce excessive smoke and creosote.

Cooking and Boiling Water on a Wood Stove

A wood stove can serve as a reliable cooking appliance during emergencies. Knowing how to prepare meals and boil water on your stove can greatly enhance your self-sufficiency.

1. **Cooking Basics:**
    - Use the flat surface of the stove to heat pots and pans.
    - Cast iron cookware is ideal for wood stove cooking due to its durability and even heat distribution.
    - Monitor food closely to avoid burning, as wood stove temperatures can be unpredictable.
2. **Baking and Roasting:**
    - Some wood stoves come with built-in ovens or attachments for baking. If your stove has this feature, familiarize yourself with its operation before an emergency.
3. **Boiling Water:**
    - Place a kettle or pot directly on the stove to boil water for drinking, cooking, or sanitation.
    - Use a lid to speed up the boiling process and conserve heat.

Staying Warm and Conserving Heat

In prolonged emergencies, conserving heat becomes critical. A wood stove can help keep your home warm, but additional measures can enhance its effectiveness:

1. **Room Insulation:**
   - Seal drafts around windows and doors using weatherstripping or draft stoppers.
   - Close off unused rooms to concentrate heat in occupied areas.
2. **Thermal Curtains and Blankets:**
   - Use heavy curtains or thermal blankets to insulate windows and trap heat inside.
3. **Positioning:**
   - Arrange furniture and sleeping areas closer to the stove for maximum warmth.
   - Avoid placing flammable items, such as blankets or furniture, too close to the stove.
4. **Layered Clothing:**
   - Dress in multiple layers, including thermal undergarments, to retain body heat.
   - Use hats, gloves, and socks to stay warm without

overburdening the stove.

Emergency Kits for Wood Stove Users

Creating a dedicated emergency kit tailored to wood stove use can streamline your response during crises. Include the following items:

1. **Fire-Starting Supplies:**
   - Waterproof matches, lighters, and fire starter logs.
2. **Backup Fuel:**
   - A reserve supply of seasoned wood or compressed wood bricks.
3. **Cooking Tools:**
   - Cast iron cookware, utensils, and heat-resistant potholders.
4. **Lighting:**
   - Candles, lanterns, and flashlights to supplement the stove's light.
5. **First Aid Kit:**
   - Include supplies for treating burns or minor injuries

related to stove use.
6. **Documentation:**
    - Manuals for the stove, carbon monoxide detectors, and any other emergency appliances.

Preparing for Extended Emergencies

In the event of prolonged power outages or natural disasters, additional preparations may be necessary to sustain wood stove use:

1. **Long-Term Fuel Storage:**
    - Plan for a fuel supply that can last several weeks or months, depending on your region and climate.
2. **Water Supply:**
    - Store bottled water or have a plan for boiling water on the stove to ensure access to safe drinking water.
3. **Emergency Communication:**
    - Keep a battery-powered radio or satellite phone to stay informed about weather updates and rescue efforts.
4. **Backup Heating Options:**

- Have additional heating methods, such as propane heaters or thermal blankets, as contingencies.

Adapting to Different Emergency Scenarios

Wood stove users should tailor their preparedness plans to the specific types of emergencies they are most likely to encounter:

1. **Winter Storms:**
    - Prioritize insulation and fuel storage to combat freezing temperatures.
    - Monitor chimney vents for ice blockages.

2. **Floods:**
    - Elevate wood supplies and stove accessories to protect them from water damage.
    - Ensure the stove's location is above flood-prone areas of the home.

3. **Earthquakes:**
    - Secure the stove and chimney to prevent tipping or structural damage during seismic activity.

4. **Hurricanes and High Winds:**

- Reinforce the chimney and ensure the stove's flue system is securely attached to avoid wind-related damage.

By preparing your wood stove and your home for emergencies, you can ensure a safer, more resilient response to any crisis.

Chapter 11: Stories from Warm, Healthy Homes

Homes are more than just buildings; they are spaces where life unfolds, memories are made, and comfort is sought. For families, couples, and individuals alike, the warmth of a wood stove often becomes the heart of the home—a source of not just heat, but also connection and well-being. In this chapter, we delve into real-life stories of people who have embraced wood stoves as part of their journey to creating warm, healthy, and harmonious homes. These accounts highlight the transformative power of combining traditional heating methods with modern strategies for health and sustainability.

A Family's Retreat: The Wilsons' Cabin in the Woods

For the Wilson family, weekends meant escaping the hustle and bustle of city life for the serene beauty of their log cabin nestled in the Appalachian Mountains. The centerpiece of their retreat was a cast-iron wood stove that radiated warmth throughout the space.

1. **Transformative Comfort:**
    - The Wilsons initially chose the wood stove for its efficiency and rustic appeal. Over time, they realized it brought their family closer together. On chilly nights, they would gather around the fire to play board games, read stories, or simply enjoy each other's company.
2. **Learning Curve:**
    - Adjusting to wood stove living wasn't without its challenges. They quickly learned the importance of

using seasoned wood to minimize smoke and maximize heat. The children even helped split and stack firewood, turning chores into cherished family moments.

3. **Healthier Air Quality:**
    - Concerned about indoor air pollution, the Wilsons installed an air purifier with a HEPA filter and ensured proper chimney maintenance. This balance of traditional and modern practices kept their cabin's air clean and their lungs clear.

An Artist's Sanctuary: Mia's Studio Haven

Mia, a ceramic artist based in Vermont, transformed an old barn into her dream studio. The space was expansive, rustic, and perfect for her craft, but the frigid New England winters posed a significant challenge. Enter the wood stove, a solution that not only provided heat but also inspired her creativity.

1. **Warmth Fuels Creativity:**
    - Mia described her wood stove as a muse. The

crackling fire became the soundtrack to her creative process, and the stove's steady warmth allowed her to work comfortably for hours, even during snowstorms.

2. **Eco-Friendly Practices:**
   - Committed to sustainability, Mia sourced her firewood from local, responsibly managed forests. She also used the stove's heat to dry her clay sculptures, integrating its functionality into her artistic workflow.
3. **A Community Connection:**
   - Neighbors often dropped by to share surplus wood or enjoy a cup of tea by the fire. The stove became a focal point for building relationships, illustrating the communal warmth it provided.

Healing with Heat: James' Journey to Respiratory Health

James, a retired teacher living in Colorado, struggled with asthma for most of his life. He was initially hesitant about installing a wood stove, fearing it might worsen his condition. However, careful planning turned his wood stove into a cornerstone of his respiratory health.

1. **Prioritizing Air Quality:**
    - James invested in an EPA-certified wood stove designed to burn cleanly and efficiently. He paired it with an advanced air quality monitor that alerted him to any spikes in particulates.
2. **Portable Nebulizer for Relief:**
    - During dry winter months, James kept a portable nebulizer on hand to manage any respiratory flare-ups. The combination of controlled heating and targeted respiratory care improved his symptoms significantly.
3. **Empowerment Through Knowledge:**
    - By educating himself about wood stove maintenance and air purification, James felt more in control of his environment. His home became a place of healing, not harm.

A Sustainable Lifestyle: The Garcias' Off-Grid Homestead

For Maria and Enrique Garcia, sustainability wasn't just a buzzword—it was a way of life. Their off-grid homestead in New Mexico relied entirely on renewable resources, with a wood stove playing a pivotal role in their energy strategy.

1. **Efficient Heating:**
    - The Garcias chose a high-efficiency wood stove that could heat their entire home with minimal fuel. The stove's secondary combustion system ensured clean burning, aligning with their eco-conscious values.
2. **Multipurpose Functionality:**
    - Beyond heating, the stove served as a cooking appliance and even a water heater. Maria often simmered stews on the stove's surface while Enrique used its heat to warm water for washing.
3. **Family Bonding:**
    - On cold evenings, the family gathered around the stove for storytelling sessions, reinforcing their connection to each other and to the rhythms of the natural world.

Urban Warmth: Sarah's City Apartment

Living in a historic rowhouse in Boston, Sarah faced unique challenges. Her apartment's central heating was inconsistent, and utility bills were steep. Installing a small, modern wood stove transformed her urban living experience.

1. **Compact Yet Powerful:**
   - Sarah's stove was designed for small spaces, offering efficient heating without overwhelming her apartment. Its sleek design complemented her modern aesthetic.
2. **Balancing Tradition and Innovation:**
   - She embraced a hybrid approach by combining the wood stove with a smart thermostat for her central heating system. This allowed her to maintain a consistent temperature while reducing energy costs.
3. **A Cozy Retreat:**
   - The stove created a cozy ambiance that made her apartment feel like a sanctuary. Friends often commented on the unique warmth and charm it added to her home.

Generational Wisdom: The Robinson Family Farm

The Robinson family had used wood stoves for generations on

their farm in rural Virginia. For them, the stove was more than just a heating appliance; it was a symbol of resilience and tradition.

1. **Passing Down Knowledge:**
    - Grandpa Robinson taught his grandchildren how to split wood, stack it properly, and start a fire with minimal effort. These lessons became cherished family rituals.
2. **Adapting to Modern Needs:**
    - While the original farmhouse stove was replaced with a modern EPA-certified model, the family continued to honor traditional practices, like using the stove's heat to dry herbs and preserve produce.
3. **A Gathering Place:**
    - The wood stove remained the heart of the home, where family and neighbors congregated during holidays, snowstorms, and quiet evenings.

Finding Balance: Emily's Journey to Holistic Wellness

Emily, a yoga instructor in Oregon, believed that her home should reflect her commitment to health and balance. Installing a wood stove became an integral part of her holistic approach to wellness.

1. **Mindful Heating:**
   - For Emily, tending to the fire was a meditative practice. Splitting wood, stacking logs, and lighting the stove became rituals that grounded her in the present moment.
2. **Creating a Healing Space:**
   - The stove's gentle warmth enhanced her yoga and meditation practice, creating an environment conducive to relaxation and self-care.
3. **Natural Aromatherapy:**
   - Emily occasionally added dried lavender or eucalyptus to the fire, filling her home with soothing scents that complemented her wellness routine.

These stories illustrate the diverse ways in which wood stoves contribute to the warmth, health, and harmony of a home. From providing physical comfort to fostering emotional connections, the humble wood stove transcends its utilitarian purpose, becoming a source of inspiration and transformation. Each narrative highlights the adaptability and enduring appeal of wood stoves in creating spaces that nurture both body and soul.

Chapter 12: Frequently Asked Questions

As more people turn to wood stoves for their heating, cooking, and ambiance needs, questions about their operation, safety, and maintenance inevitably arise. In this chapter, we address the most common queries about wood stoves, providing clear, practical answers based on expert knowledge and real-world experience. Whether you're a first-time user or a seasoned wood stove owner, this FAQ section is designed to resolve doubts and offer actionable advice.

---

1. What type of wood stove should I buy?

Answer: The type of wood stove you should buy depends on your specific needs:

- **Heating Needs:** If you're using the stove as your primary heat source, choose a model with higher BTU output suitable for your home's square footage.
- **Space:** For smaller spaces, compact or wall-mounted models are ideal. For larger areas, freestanding stoves with higher efficiency ratings work best.
- **Catalytic vs. Non-Catalytic:** Catalytic stoves offer higher efficiency and longer burn times but require more maintenance. Non-catalytic stoves are simpler to operate and maintain.
- **Style and Design:** Modern stoves offer a variety of aesthetic options, from rustic cast iron to sleek steel models. Select one that complements your home's decor.

---

2. How often should I clean my wood stove?

Answer: Cleaning frequency depends on how often you use your stove:

- **Daily Use:** Clean the ash pan every one to three days and check the glass for soot buildup.

- **Monthly Maintenance:** Inspect and clean the baffle and air intake vents to ensure proper airflow.
- **Seasonal Cleaning:** Before the heating season begins, thoroughly clean the firebox, stovepipe, and chimney to remove creosote and debris.

Regular maintenance not only improves efficiency but also reduces the risk of chimney fires.

3. Can I burn any type of wood in my stove?

Answer: No, not all wood is suitable for burning in a wood stove. Here's what to consider:

- **Seasoned Wood:** Always burn wood with a moisture content below 20%. Seasoned hardwoods like oak, maple, and hickory are best for long-lasting heat.
- **Avoid Green Wood:** Green or unseasoned wood produces excessive smoke and creosote.
- **Prohibited Materials:** Never burn treated wood, painted wood, driftwood, or household waste, as these can release toxic fumes and damage your stove.

4. How do I prevent creosote buildup?

Answer: Creosote is a byproduct of burning wood and can accumulate in the chimney, posing a fire hazard. To minimize buildup:

- Burn only seasoned wood.
- Maintain a steady, hot fire to promote complete combustion.
- Avoid smoldering fires, which produce more creosote.
- Schedule annual chimney cleanings by a certified professional.
- Use creosote-removal logs periodically to reduce buildup.

---

5. What should I do if smoke enters the room?

Answer: Smoke backdrafts can occur for several reasons:

- **Cold Chimney:** Preheat the chimney by burning a small amount of newspaper before starting your fire.
- **Blockages:** Check for obstructions in the chimney or flue.
- **Negative Air Pressure:** Open a window slightly to balance indoor and outdoor air pressure.
- **Improper Draft:** Ensure the stove's damper is fully open

when starting a fire.

If the issue persists, consult a professional to inspect the system.

---

6. How can I improve the efficiency of my wood stove?

Answer: Maximize your stove's efficiency with these tips:

- Use properly seasoned hardwoods for optimal heat output.
- Adjust air vents to regulate combustion.
- Keep the stove's glass and air intake clean for better airflow.
- Install a stove thermometer to monitor and maintain ideal temperatures (between 300°F and 500°F).
- Consider upgrading to an EPA-certified stove for improved efficiency.

---

7. Is it safe to leave my wood stove burning overnight?

Answer: Yes, but only if proper precautions are taken:

- Build a slow-burning fire using large, dense logs.
- Close the air vents partially to reduce airflow and slow

combustion.
- Ensure the firebox is clean and the chimney is unobstructed.
- Never leave flammable materials near the stove.
- Install smoke and carbon monoxide detectors to alert you to any hazards.

8. How do I properly store firewood?

Answer: Proper storage ensures your wood remains dry and ready for burning:

- Stack wood off the ground using pallets or a wood rack.
- Cover the top of the stack with a tarp, leaving the sides exposed for ventilation.
- Store wood in a sunny, breezy location to speed up drying.
- Rotate your stock, using older wood first to prevent it from rotting.

9. Are wood stoves environmentally friendly?

Answer: Modern wood stoves can be an eco-friendly heating option when used correctly:

- **Carbon Neutral:** Burning wood releases the same amount of carbon dioxide as it would during natural decomposition.
- **Efficient Models:** EPA-certified stoves reduce emissions by up to 90% compared to older models.
- **Sustainable Sourcing:** Use wood from responsibly managed forests to minimize environmental impact.

---

10. Can I install a wood stove myself?

Answer: While it's possible for experienced DIYers to install a wood stove, professional installation is strongly recommended:

- Professionals ensure the system meets local building codes and safety standards.
- Improper installation can lead to chimney fires, carbon monoxide leaks, or structural damage.
- Hiring a certified installer often preserves the stove's warranty.

---

11. What is the lifespan of a wood stove?

Answer: A well-maintained wood stove can last 15-30 years or more. Factors affecting longevity include:

- The quality of the stove's construction.
- Regular maintenance and cleaning.
- Proper use of seasoned wood and adherence to manufacturer guidelines.

---

12. Do I need a backup heating system?

Answer: While wood stoves are reliable, having a backup heating system is recommended for emergencies, such as:

- Running out of wood during extreme weather.
- Stove malfunction or chimney blockages.
- Supplementing heat in areas the stove cannot reach.

Consider electric space heaters, propane heaters, or central heating as backups.

---

13. Can I use a wood stove in an apartment or rental property?

Answer: This depends on local regulations and landlord policies:

- Many rental agreements prohibit wood stoves due to

safety concerns.
- If permitted, ensure the stove meets all safety codes and has proper ventilation.
- Consider small, portable wood stoves or alternative heating methods for compact spaces.

---

14. What should I do during a power outage?

Answer: A wood stove is an excellent heating option during power outages:

- Keep a stockpile of seasoned wood ready for emergencies.
- Use the stove's surface for cooking or boiling water.
- Ensure you have flashlights, batteries, and a fire extinguisher on hand.
- Monitor indoor air quality and ventilation during extended use.

---

15. How do I dispose of ash safely?

Answer: Proper ash disposal minimizes fire hazards:

- Use a metal container with a tight-fitting lid to store cooled ash.
- Place the container on a non-combustible surface, away from buildings.
- Once completely cool, ash can be used as garden fertilizer or disposed of with household waste, depending on local regulations.

This comprehensive FAQ section addresses the most pressing concerns about wood stove use, helping users enjoy their stoves safely and efficiently. As you continue your journey with wood stoves, these answers will serve as a reliable resource for maintaining warmth and well-being in your home.

Conclusion: Warmth, Health, and Harmony

The flickering flame of a wood stove is more than just a source of heat; it is a timeless symbol of resilience, comfort, and connection. From the crackling embers of ancient hearths to the sleek, modern designs of today, wood stoves have remained central to the human experience. In embracing a wood stove, you are not merely heating your home—you are cultivating a lifestyle that prioritizes warmth, health, and harmony.

Warmth Beyond Temperature

Warmth transcends physical comfort. It is an intangible quality that transforms a house into a home, a room into a refuge. The steady glow of a wood stove offers a kind of heat that no central heating system can replicate—a heat that penetrates not just walls but also the human spirit.

1. **Emotional Comfort:**
    - The act of tending a fire connects us to centuries

of tradition, grounding us in the present while linking us to the past.
    - The soothing sounds of crackling wood and the mesmerizing dance of flames create a meditative environment that calms the mind and nurtures the soul.
2. **Social Connection:**
    - Wood stoves naturally draw people together, fostering moments of togetherness and shared experience. Family game nights, intimate conversations, or solitary moments of reflection all gain a deeper dimension in the glow of a fire.
3. **Resilience and Self-Sufficiency:**
    - A wood stove empowers you to take control of your heating needs, fostering independence from external energy sources. This sense of resilience is particularly rewarding during power outages or extreme weather events.

Health in Every Breath

Health is a cornerstone of a well-lived life, and the choices we make in our homes have profound effects on our well-being. Wood stoves, when used thoughtfully, can support both physical and mental health.

1. **Respiratory Wellness:**
    - Modern EPA-certified wood stoves produce fewer emissions and burn more efficiently, reducing indoor air pollution.
    - Pairing a wood stove with proper ventilation, air purifiers, and the use of seasoned wood minimizes risks and promotes clean air.
2. **Therapeutic Benefits:**
    - The radiant heat from a wood stove can soothe sore muscles and provide relief for conditions exacerbated by cold weather, such as arthritis.
    - The ritual of fire-starting and tending offers a mindful practice that reduces stress and enhances mental clarity.
3. **Environmental Health:**
    - By using sustainably sourced wood and maintaining a clean-burning stove, you contribute to a healthier planet. Carbon-neutral heating reduces your carbon footprint, aligning your lifestyle with eco-conscious values.

Harmony in Home and Hearth

Harmony is the delicate balance that allows all elements of life to coexist peacefully. In a home with a wood stove, harmony is found in the interplay between tradition and modernity, sustainability and convenience, warmth and health.

1. **Design and Function:**
   - Today's wood stoves marry functionality with aesthetics, offering designs that complement diverse architectural styles. Whether your home is rustic, minimalist, or contemporary, there's a stove to suit your vision.
2. **Integration of Technology:**
   - Smart features like remote temperature controls and real-time air quality monitors blend seamlessly with the timeless appeal of wood-burning heat.
   - Advances in stove technology, such as catalytic converters and secondary combustion systems, ensure optimal performance with minimal environmental impact.

3. **Sustainable Living:**
    - The use of renewable energy sources like wood fosters harmony with nature, aligning your home's energy consumption with the cycles of the earth.
    - By choosing sustainable practices, such as local wood sourcing and efficient heating methods, you contribute to a global effort toward environmental stewardship.

Stories That Illuminate

The value of a wood stove extends beyond practicality, finding its true expression in the stories it creates. Each chapter of this book has touched upon narratives that illustrate how wood stoves shape lives, from cozy family moments to inspiring tales of resilience.

1. **Personal Growth:**
    - Learning to operate and maintain a wood stove fosters skills like patience, resourcefulness, and problem-solving.
    - The act of splitting, stacking, and burning wood connects users to physical work and nature in a

way few other heating methods can.
2. **Cultural Significance:**
    - In many cultures, the hearth symbolizes the heart of the home, a place where traditions are passed down and community is built.
    - Stories of wood stoves often intertwine with larger narratives of sustainability, craftsmanship, and heritage.

Practical Wisdom for the Future

As you embrace the benefits of a wood stove, the knowledge and practices shared in this book equip you to navigate challenges and maximize rewards. Here are key takeaways:

1. **Preparation:**
    - Regular maintenance, proper installation, and a well-stocked woodpile are foundational to safe and efficient stove use.
2. **Mindful Usage:**
    - Burning the right type of wood, monitoring air quality, and maintaining proper ventilation ensure that your wood stove supports your health and

safety.
3. **Adaptability:**
    - Whether you're using a wood stove as a primary heat source, an emergency backup, or an aesthetic addition, flexibility is key to integrating it into your lifestyle.
4. **Continuous Learning:**
    - As technology advances and environmental concerns evolve, staying informed ensures that you remain a responsible and effective wood stove user.

The Power of Choice

Ultimately, the decision to incorporate a wood stove into your home is a profound one. It reflects a commitment to:

- **Personal Comfort:** Creating a sanctuary of warmth and tranquility.
- **Health and Well-Being:** Prioritizing clean air and sustainable practices.
- **Environmental Responsibility:** Aligning with renewable energy solutions that respect the planet.

By choosing a wood stove, you are embracing a lifestyle that is both timeless and forward-thinking. It is a choice that honors the past, enriches the present, and contributes to a sustainable future. The stories, insights, and practical advice in this book are designed to guide you on that journey, ensuring that your home remains a haven of warmth, health, and harmony.

www.ingramcontent.com/pod-product-compliance
Lightning Source LLC
Chambersburg PA
CBHW050304230526
45471CB00005B/2016